Water
112

咀嚼成汤
Chew Your Soup

Gunter Pauli

[比] 冈特·鲍利　著

[哥伦] 凯瑟琳娜·巴赫　绘

郭光普　译

上海远东出版社

丛书编委会

主　任：田成川

副主任：闫世东　林　玉

委　员：李原原　祝真旭　曾红鹰　靳增江　史国鹏
　　　　梁雅丽　孟小红　郑循如　陈　卫　任泽林
　　　　薛　梅　朱智翔　柳志清　冯　缨　齐晓江
　　　　朱习文　毕春萍　彭　勇

特别感谢以下热心人士对童书工作的支持：

匡志强　宋小华　解　东　厉　云　李　婧　庞英元
李　阳　梁婧婧　刘　丹　冯家宝　熊彩虹　罗淑怡
旷　婉　王靖雯　廖清州　王怡然　王　征　邵　杰
陈强林　陈　果　罗　佳　闫　艳　谢　露　张修博
陈梦竹　刘　灿　李　丹　郭　雯　戴　虹

目录

Contents

ZERI Learning Initiative

一只仓鼠一刻不停地咀嚼着，脸颊被咀嚼过的食物塞得鼓鼓的。一只小狐狸看着他，惊讶于仓鼠能把食物嚼成浓汤状。

　　"你一定是咀嚼冠军了。没人比你嚼得更细了！"狐狸评论道。

A hamster is chewing and chewing away, until his cheeks are bulging with chewed food. A young fox watching him, is impressed by how the hamster is able to chew his food so fine that it looks like a thick soup.

"*Y*ou must be the champion. No one chews more than you do," comments the fox.

一只仓鼠不停地咀嚼着……

A hamster is chewing away...

······像土豆泥一样。

... like mashed potatoes.

"难道你就一口吞下去而不咀嚼吗？"仓鼠问道，"要知道，对我们仓鼠来说把食物嚼碎一点很重要，要嚼得像土豆泥一样。"

"但是看起来你好像在咀嚼已经被咀嚼过的食物。"狐狸说。

"Aren't you the one who just swallows and does not chew at all?" Hamster asks him. "You know, it is important for us hamsters to chew our food into fine pieces; so fine that it becomes like mashed potatoes."

"But it looks as if you chew what has already been chewed," Fox says.

"噢，我们咀嚼不仅是把食物嚼碎，还要把食物和数百万的细菌和大量的酶混合起来，让它们在我们嘴巴里繁荣生长。"

"数百万的细菌？我还以为细菌对于我们并不好。"

"得了吧，如果没有细菌我们怎么能生存？细菌代表地球上最初的生命形式。它们已经存在数十亿年了，比我们四足行走的生物要久远得多！"仓鼠告诉狐狸。

"Well, chewing is not only about mincing food to pieces, we need to mix it with the millions of bacteria and the wealth of enzymes that are happily thriving in our mouths."

"Millions of bacteria? I thought bacteria are not good for us."

"Come on, how could we ever live without bacteria? Bacteria represent the first form of life on Earth. They have been around for billions of years – long before we starting walking on four paws," Hamster informs him.

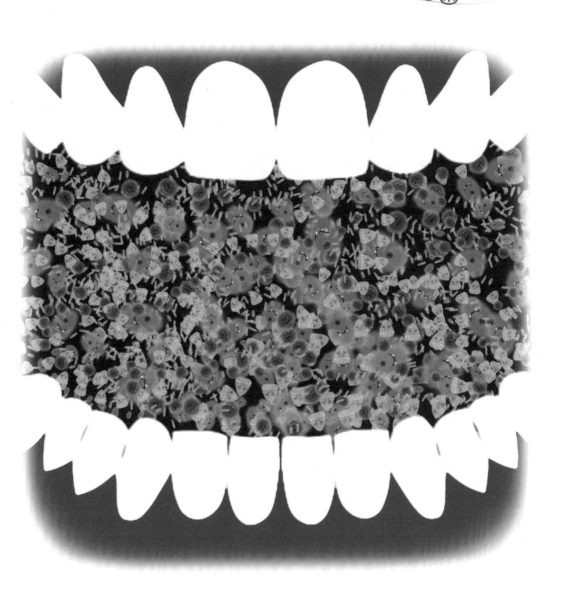

……数百万在我们嘴里繁荣生长的细菌……

… millions of bacteria that thrive in our mouths …

你在逗我吗?

Are you serious?

"你在逗我吗?" 狐狸问道。

"有人告诉我一个人嘴巴里的细菌数量要比地球上的人口总量还要多。"

"只是一个人的嘴巴里就有这么多细菌吗? 我想知道这数以亿计的细菌在嘴巴里做什么? " 狐狸疑惑道。

"Are you serious?" Fox asks.

"I have been told that a human has more bacteria in his mouth than there are people living on Earth."

"In just one person's mouth?These billions of bacteria, I wonder what they are all doing there?" Fox wants to know.

"我们需要这些细菌帮我们把食物中存在的长链分子切成小段。"

"分子？分子又是啥玩意儿？是动物、植物还是细菌？"

"你不知道什么是分子吗？好吧，分子是由大量原子聚合而成的。"仓鼠告诉狐狸。

"We need them to break the long strings of food molecules into smaller pieces, you know."

"Molecules? What is a molecule again? Is it an animal, a plant or a bacterium?"

"You don't know what molecules are? Well, a molecule is a bunch of atoms that stick together," Hamster tells Fox.

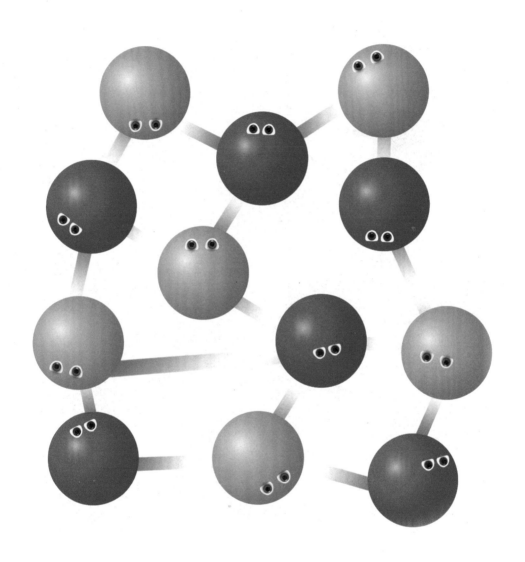

...a bunch of atoms that stick together ...

"听着，我知道你很聪明，这些你都懂，但对我来说全都是新的知识。我能认为世界上任何事物、任何人都是由分子和原子组成的吗？"

"当然！这些原子全都有自己的名字，它们是事物最小的组成单位。分子则是比原子大一点的单位，看起来就像是神奇的结构式。"仓鼠解释道。

"Look, I know you are smart and know all about this, but this is all new to me. Do I get it right that everything and everyone is made up of atoms and molecules?"

"Exactly, those atoms, all with different names, are the smallest building blocks. And molecules are the larger building blocks that look like magical formulas," Hamster explains.

"我设想是许多小的零件连接在一起组成一个大的结构。所以，这是不是你必须充分咀嚼的原因——把这些紧密连在一起的结构分开？"狐狸问道。

　　"是的，你说得对，你的确很聪明！当我们把食物吞进肚子里，我们应该确保这些食物已经很小了，这样营养才能更容易进入我们的血液中。"

"And many small ones put together to make a big one, I imagine. So is that then why you have to chew properly – to take apart what has been carefully put together?" Fox asks.

"Yes, you've got it. You really are smart! When we put food into our stomach, we should make sure it is in such tiny bits that it can get into our blood easily."

......细小的微粒更容易进入我们的血液。

... tiny bits to get into our blood easily.

......食物的一部分会留在肚子上......

...some ends up on our belly...

"你没在开玩笑吧？我们吃的大块食物最后会变成小分子进入血液中，然后增强我们的体质？"

"很不幸，食物的一部分会留在肚子上。"仓鼠说道。

"Are you serious, the big pieces of food we chew end up as tiny pieces in our blood, building our bodies?"

"Unfortunately, some of it ends up on our belly," Hamster says.

"你的意思是：我们肚子里的食物？"

"不，你好好听我讲。当我们吃得太多时，最终就会转化为腹部脂肪。跟我们是否咀嚼恰当无关，而是因为我们选择了错误的食物！"

……这仅仅是开始！……

"You mean: food in our belly?"

"No, you heard me correctly. When we eat too much, we end up having to carry that extra weight as belly fat! Not because of chewing properly or not – but because we choose the wrong foods to chew on!"

... AND IT HAS ONLY JUST BEGUN! ...

AND IT HAS ONLY JUST BEGUN! ...

Did You Know?

你知道吗?

Chewing food into small pieces permits food particles to have a much larger contact surface, exposing it to saliva and bacteria for longer periods. Food exposed to saliva is well lubricated and will ease swallowing.

把食物嚼碎使得食物颗粒和消化液接触更充分，可以长时间接触唾液和细菌。混合了唾液的食物颗粒会变得很光滑，更容易下咽。

If food is not finely chewed, parts of it will be incompletely digested. This leads to a loss of nutrition, and the undigested food becomes fodder for bacteria in the colon, which leads to flatulence.

如果食物没有被咀嚼得很充分，有些部分就不会被完全消化。这不仅仅会导致营养流失，这些没被消化的食物还会成为结肠细菌的食物，并导致肠胃气胀。

Chewing sends a stream of messages to the digestive system. The longer one sees, smells and tastes food, the more the nervous system is prompted to optimise the process of metabolism.

咀嚼会给消化系统传递一连串信息。一个人对食物看、闻和品尝的时间越久，就越能促进神经系统优化新陈代谢过程。

能量

The more you chew, the longer it takes to finish a meal, and the less you will eat, avoiding weight gain. Digestion is a demanding task, requiring a lot of energy. So the longer you chew, the more energy the body has for other important tasks.

咀嚼得越细，吃饭所花费的时间就越长，你吃得就越少，从而避免了长胖。消化是一个高要求的任务，会消耗大量能量。所以咀嚼的时间越长，身体分给其他重要任务的能量就越多。

The jawbones in which your teeth are set get a "workout" when you chew, helping to keep them strong. Saliva produced while chewing, clears food particles from your mouth and washes away bacteria so there will be less plaque build-up and tooth decay.

在你咀嚼的时候，牙齿所在的下颚骨会得到锻炼从而保持强健。在咀嚼的时候还会产生唾液，把嘴巴里的食物碎屑清理干净并把细菌带走，这样会减少牙菌斑滋生和蛀牙产生。

Hamsters have very poor eyesight. They leave a scent trail as a guide to get home again. Hamster's teeth grow continually. Chewing keeps their teeth short. A mother hamster is able to put all her young in her cheek pouches and carry them to safety.

仓鼠的视力非常差。它们根据气味的痕迹来寻找回家的路。仓鼠的牙齿会不断生长，咀嚼能使它们的牙齿保持在比较短的状态。仓鼠妈妈能把它所有的孩子都放进自己的颊囊里并把它们带到安全的地方。

Hamsters have spacious cheek pouches to carry food for underground storage in their burrows. When full, their cheek pouches can double or even triple the size of their heads. A chipmunk's head can even be as large as its body when its cheek pouches are full.

仓鼠有特别大的颊囊，能把食物带回洞穴中地下储藏。它们的颊囊装满食物的时候比它们的头大两倍甚至三倍。花栗鼠的颊囊装满的时候，头甚至能和身体一样大。

Foxes do not chew their food. Their teeth are sharp, and are used to cut food into smaller chunks.

狐狸不咀嚼食物。它们的牙齿很锋利，能把食物撕成碎片。

Do you eat fast or do you eat slowly? What are the implications for you?

你吃东西速度是慢还是快？这对你有什么影响？

What about organising a daily workout for your teeth and bones by adding extra chewing time?

平日里通过多咀嚼来锻炼你的牙齿和骨骼，如何？

Are the bacteria in your mouth good for you and your digestion?

你嘴里的细菌对你和消化有好处吗？

How logical is it that before food can build your body and provide energy, it must be turned into the smallest pieces possible?

食物在进入你的身体为你提供能量之前必须先变为尽量小的颗粒。你觉得这是怎样的逻辑？

Do It Yourself!

自己动手!

Take a bite of bread and chew it, and then chew it some more, and some more. Keep on chewing – do not swallow! As you do this, pay attention to how the flavour of the bread changes.

What is happening is that the bread, which is made up of starch (a long molecule chain in a carbohydrate form, which your body cannot easily absorb), is broken up into small sugar molecules – called glucose – by an enzyme in your saliva. Experience for yourself how a lot of chewing turns bread into a sweet product.

Now try to swallow some bread without chewing it. Compare this to swallowing bread after you have chewed it at least 50 times.

咬一口面包并咀嚼，多咀嚼一会儿，再咀嚼一会儿。继续咀嚼——不要吞下去！在你做这些时，注意面包的味道是如何变化的。

面包是由淀粉（一种长分子链的碳水化合物，不容易被身体吸收）组成的，被唾液中的酶转化为小分子糖——葡萄糖。自己感受一下，多嚼一嚼，面包就会变甜。

现在试着不咀嚼吞下一些面包。与咀嚼了至少 50 次之后再吞下去的感觉比较一下。

27

学科知识
Academic Knowledge

生物学	仓鼠是啮齿动物；夜行（夜间活跃，白天睡觉）；冬眠；狐狸生活在地下洞穴里；狐狸不咀嚼食物，使用裂齿把肉撕成块；新生的狐狸幼崽在开始的两周是不睁眼的；狐狸吃蠕虫、蜘蛛、浆果、老鼠和鸟类；食管需要比较湿润的食物，这样吞下时可以减少阻力。
化 学	唾液中的α-淀粉酶能分解碳水化合物，舌脂酶能够分解脂肪；咀嚼时人体会释放胆囊收缩素、生长激素抑制素和神经降压素；对胰腺来说，咀嚼信号使胰腺将碳酸氢盐分泌到小肠腔中；摩尔（mol）是物质的量的单位。
物 理	原子和分子的区别；狐狸有光敏感细胞（透明的反光色素层），所以有夜视能力；最小的分子是氢分子（H_2）；一个原子的直径范围为0.1到0.5纳米。
工程学	仓鼠生活在地下，在炎热的气候下，它们的洞穴能够保持凉爽，仓鼠把食物储存在其中。
经济学	饮食功能紊乱而导致的经济成本问题包括医疗保健、就业状况、就业能力、收入损失，这意味着治疗和预防均具有广泛的经济影响。
伦理学	饮食中的问题不仅仅是没咀嚼好，我们还可能误吃了各种各样难以消化的食物，这导致了肥胖症大量出现；在中世纪的欧洲，猎狐被认为是一种"运动"，但其目的是杀死狐狸，所以不能被称为一种运动。
历 史	伊索寓言可以追溯到公元前6世纪，其中有一些关于狐狸的故事，最著名的一则是《狐狸和葡萄》；《列那狐的故事》出现在公元11—12世纪；尼可罗·马基亚维利说成功的王子必须有狮子和狐狸的特征；1668年,法国文学家让·德·拉·封丹将伊索寓言改编成诗歌，其中包括一些涉及狐狸的故事；在法国作家安托万·德·圣-埃克苏佩里的著作《小王子》中，狐狸告诉我们友谊的真正价值。
地 理	仓鼠在叙利亚被首次发现；它们住在大草原、沙丘和沙漠边缘；狐狸出现在从沙漠到北极的生态区中。
数 学	用组成分子的各种原子的重量乘以原子个数，并求它们的和，来计算一个分子的重量。
生活方式	在野外，一只狐狸需要20平方千米的空间才能生存下去；适应城市地区的狐狸只需要0.1平方千米，因为人们挥霍浪费的生活方式给它们提供了足够的食物。
社会学	凯尔特文化认为狐狸是明智的，像向导一样，还形成了谚语"像狐狸一样聪明"；美国本土文化认为狐狸是骗子。
心理学	我们害怕任何我们看不见的东西——如即使细菌是生命不可或缺的，而且许多细菌是有益的，人们还是容易认为细菌是有害的或危险的。
系统论	在我们开始考虑环境健康之前，我们需要花费时间和精力来改善我们的身体健康。

情感智慧
Emotional Intelligence

狐 狸

狐狸认识到仓鼠的独特能力。他坦诚地和仓鼠进行交流，甚至可以逗他笑，这表明他们之间有着良好的关系。狐狸质疑细菌的作用。听了仓鼠介绍后，狐狸对人体中细菌的数量非常惊讶。狐狸承认自己对细菌的作用及对原子和分子等知识的缺乏，并主动询问更多的信息。他希望了解咀嚼的作用，食物如何从大块转变成微小的颗粒，这些微粒如何通过血液循环分布到全身各处。

仓 鼠

仓鼠有爱挑战的性格，告诉了爱学习的狐狸关于仓鼠是如何处理食物的信息。仓鼠很乐意解释这个过程和逻辑，并提供了细菌的作用的历史背景。然而，狐狸的问题确实为仓鼠提供了一个分享他的知识的机会。他很惊讶地发现，被认为聪明的狐狸知道得少之又少。他显得很失望，因为最基本的概念也要给予说明。尽管如此，仓鼠仍能继续探讨，并在关于选择正确食物的观点上结束了对话。

艺术
The Arts

品茶是伟大的艺术之一！当你品尝并允许它停留在你的嘴里和舌头周围的时候，你才能发现茶内在的真正秘密。此时茶与唾液混合会产生独特的味觉体验，你将感受到苦涩和甜蜜。只有当茶接触到舌头上的味蕾时才会品尝到它的味道，因为这些味道都不能通过嗅觉来感受。起作用的不仅是味道，还有口感。在品茶的艺术中，你需要区分出来品尝的开始、中间和结束部分之间的不同。

思维拓展
Systems: Making the Connections

我们已经不再去感受我们的食物的数量、质量、颜色、形状、香味、味道和浓度。在吃饭的时候，人们一直在看智能手机或其他电子设备，不断地下载和"嘟嘟"响，使我们的注意力从餐盘转移开。我们狼吞虎咽地吃完饭，然后期望我们的胃去做剩下的事情。我们没有足够的时间来吃饭。我们似乎不再享受家人和朋友的陪伴，我们当然也没有足够的咀嚼。咀嚼是健康饮食的关键。如果我们不花时间来嚼碎食物，将其与不同的酶混合，不仅会伤害我们的牙齿，还把压力留给了吞咽过程。当我们正确咀嚼时，食道的工作会更容易，也会促进消化过程。我们的大脑需要大约20分钟的时间来意识到胃部装满了，这传达了一个明智的决定：不要吃得过多。因此，适当的咀嚼可以提供一个缓冲功能。咀嚼用到了所有牙齿，它们连接到身体的每一个经络，激活身体，并准备好胰脏的功能。这将确保酸和碱液的适当混合物准备好并应用于顺利和健康的消化和新陈代谢。如果食物没有消化好，肠道会超负荷工作。这就导致了有害细菌和真菌由于有食物来源而不断生长。这些都会影响我们的体重、健康和免疫系统的能力。我们的身体有一个复杂的系统，当我们触发它工作的时候，它运行得非常好。重要的是我们要知道身体是如何运作的，并要照顾好我们的身体健康。如果我们没有知识，也不关心我们自己具体的生活，我们怎么能理解和保护我们所生活的环境呢？

动手能力
Capacity to Implement

咀嚼的艺术。大脑大约需要20分钟才能意识到胃里装满了，然后发送相应的消息来有意识地停止进食。由此可知，在吃东西的时候慢下来并适当地咀嚼至少有以下四种好处：（1）通过改善消化水平来控制体重；（2）锻炼牙龈、牙齿和骨骼；（3）促进必要的营养物质的吸收；（4）享受一种新的味道。一旦你脑中对争论有了清晰的判断，就和朋友们一起吃饭，把这个转变成餐桌上一个有趣的话题。

故事灵感来自
This Fable Is Inspired by

科瑞娜·塔尔尼策
Corina Tarnita

科瑞娜是美国新泽西州普林斯顿大学的生态学和进化生物学助理教授。她的研究重点是生物系统中复杂相互作用的动力学。虽然她主要用理论方法，但也结合了进化动力学、进化博弈论和网络理论。她是一个跨学科的研究者，作为实验和野外生态学家、分子生物学家和进化生物学家在看似不同的领域工作。她将所有的见解整合到模型和实验工作中。她最大的研究兴趣集中在细菌、昆虫和人类的社会行为，如种群结构和空间格局对进化和生态动力学，尤其对共生体的多物种网络关系中相互作用的影响。咀嚼及其效果是她的基础工作之一。

图书在版编目（CIP）数据

冈特生态童书.第四辑:修订版:全36册:汉英对照 /
(比)冈特·鲍利著;(哥伦)凯瑟琳娜·巴赫绘;
何家振等译. —上海:上海远东出版社,2023
书名原文: Gunter's Fables
ISBN 978-7-5476-1931-5

Ⅰ.①冈… Ⅱ.①冈… ②凯… ③何… Ⅲ.①生态环
境–环境保护–儿童读物—汉、英 Ⅳ.①X171.1-49

中国国家版本馆CIP数据核字(2023)第120983号
著作权合同登记号图字09-2023-0612号

策　　划　张　蓉
责任编辑　张君钦
封面设计　魏　来　李　廉

冈特生态童书
咀嚼成汤
[比]冈特·鲍利　著
[哥伦]凯瑟琳娜·巴赫　绘

郭光普　译

记得要和身边的小朋友分享环保知识哦！
八喜冰淇淋祝你成为环保小使者！